COSMIC SIGNATURES

Tracing the Creator Through Science

MM Williams

Copyright © 2024 MM Williams

All rights reserved.

Printed in the United States of America

For information and permission to reproduce selections from this book, write to MMWilliamsPoet@Outlook.com

ISBN: 9798343722482

DISCLAIMER

The content presented in this book represents a synthesis of scientific theories, philosophical arguments, and spiritual interpretations. While efforts have been made to explain these ideas in an accessible and balanced manner, it is important to recognize that many of the topics discussed—such as the existence of a creator, the origin of the universe, and the nature of consciousness—remain matters of ongoing debate and inquiry across various disciplines.

The interpretations and conclusions presented herein are not intended to represent definitive answers, nor do they necessarily reflect the views of all experts in the fields of science, philosophy, or theology. Readers are encouraged to explore these topics further and to engage with a variety of perspectives to form their own informed opinions.

This book is designed to inspire curiosity and reflection, but it should not be taken as a substitute for rigorous academic study or professional advice on scientific or spiritual matters. Additionally, where references are made to specific scientific theories or philosophical ideas, these are intended as simplified explanations and should not be seen as comprehensive treatments of complex subjects.

The exploration of existential questions is deeply personal, and readers are encouraged to approach the material with an open mind while also critically considering the ideas presented.

ABOUT THE AUTHOR

MM Williams is a master of multiple genres, seamlessly traversing the realms of poetry, suspense thrillers, science fiction, philosophy, and science. With a prolific body of work, Williams is as adept at weaving lyrical, romantic verses as he is at crafting edge-of-your-seat thrillers and profound philosophical explorations. His poetry collections are lyrical, brimming with dense, vivid imagery and emotional depth, blending existential questions with themes of spirituality and romance. In his suspense thrillers, Williams unravels complex geopolitical, psychological and dystopian plots, laced with espionage and high-stakes intrigue tinged with mystery, delivered with razor-sharp precision. His science fiction is distinguished by its rich interweaving of historical facts, Greek mythology, and philosophical inquiry, creating worlds that challenge both the mind and imagination.

As a thinker and philosopher, Williams delves into the mysteries of consciousness, the mind-body connection, and holistic well-being, offering readers deep insights on self-transformation and the nature of existence. His science writing, driven by a quest to understand the cosmos, time, and the fabric of reality, makes the wonders of the universe both accessible and awe-inspiring, illuminating complex concepts with clarity and passion.

Williams' versatility as a writer lies in his unique ability to blend the emotive power of poetry, the pulse-pounding thrill of suspense, the depth of philosophical reflection, and the boundless curiosity of scientific exploration. His work not only challenges and captivates but also profoundly moves readers, resonating across the entire spectrum of human experience.

FOREWORD

In our quest to understand the mysteries of existence, we have always been captivated by the profound questions that arise from the very fabric of reality. From ancient myths whispered around the fire to the intricate theories of modern science, humanity has long sought to unravel the enigma of our origins. Where did we come from? Why does the universe exist as it does? Is there an underlying purpose, a creator, or are we the product of chance in a self-organizing cosmos?

These questions have shaped the course of human thought for millennia. Philosophers, theologians, and scientists have all, in their own ways, grappled with these timeless inquiries. In this book, we delve into some of the most compelling evidence and arguments drawn from both science and metaphysics to explore the possibility of a creator—an intelligence behind the grandeur of the universe.

The journey laid out in these chapters reflects a unique fusion of disciplines: the precision of physics, the poetic wonder of metaphysics, and the spiritual wisdom of both Eastern and Western traditions. We will consider how the fine-tuned laws of nature, the complexity of life, the mysteries of consciousness, and the abstract beauty of mathematics might point us toward deeper truths about existence itself. Each chapter takes you one step closer to the heart of the question: Is there something beyond what we can see and measure?

Yet, this book is not merely an intellectual exercise. It is an invitation to reflect, to slow down, and to connect with the awe that permeates the universe and our place within it. Whether you approach this journey as a scientist, a seeker, or simply a curious mind, my hope is that these pages inspire you to engage with the profound mystery of existence—no matter where you stand on the question of a creator.

The answers may remain elusive, but in the search itself, there is beauty. Science and spirituality need not be opposing forces; rather, they are two complementary paths leading us toward the same ultimate wonder. In the end, this book does not claim to provide definitive conclusions, but it encourages you to embrace the questions — and, in doing so, discover your own sense of meaning in the cosmos.

Welcome to a journey through the infinite — the known and the unknown, the measurable and the transcendent. Let us explore the possibilities together.

PREFACE

This book began as a simple question: Why does the universe exist in the way that it does? Like many who contemplate the vastness of space, the precision of the natural laws, and the intricacies of life itself, I found myself drawn to the idea that behind this profound order, there may lie a deeper intelligence, a creator, or perhaps a force beyond our current understanding.

As a writer and thinker with a passion for both science and spirituality, I felt compelled to explore these questions not from a place of rigid certainty, but from a place of curiosity and wonder. The more I delved into the scientific discoveries of our time — ranging from the Big Bang theory to the remarkable structure of DNA — the more I saw parallels between the insights of modern science and the timeless wisdom of ancient philosophies.

The journey of writing this book has been one of exploration — bridging the worlds of physics and metaphysics, cosmology and consciousness, and science and spirituality. I have attempted to present the arguments for and against a creator's existence in a way that is both accessible and profound, allowing readers of all backgrounds to engage with these complex ideas. The goal is not to provide a definitive answer to the age-old question of creation, but to present the evidence and theories in a way that invites thoughtful reflection.

Each chapter explores a different dimension of this grand inquiry — from the fine-tuning of the universe to the mysteries of quantum mechanics, from the precise laws of nature to the beauty of mathematical order. As you read, you will encounter the arguments of great minds across history, from classical philosophers like Aquinas and Kant to contemporary scientists and thinkers who push the boundaries of what we know.

At its heart, this book is an invitation. It is an invitation to pause, to reflect, and to embrace the wonder of existence. Whether or not you believe in a creator, the universe itself is full of mystery, and in seeking to understand it, we engage with the deepest aspects of our own nature.

I am deeply grateful to those who have inspired my thinking along the way — both the scientists who explore the limits of the physical world and the spiritual teachers who remind us of the unseen. I hope that, regardless of where you find yourself on the spectrum of belief, this book will resonate with your sense of awe and curiosity.

In the end, the question of a creator may be less about finding answers and more about deepening our relationship with the cosmos and our place within it. The search itself is a source of meaning, and it is my sincere wish that this journey will bring you closer to the mysteries of life that unite us all.

— MM Williams

"The universe is not only stranger than we imagine, it is stranger than we can imagine."
– *J.B.S. Haldane*

The Fine-Tuning of the Universe

In the vastness of the night sky, where stars flicker like distant memories and the cool breath of the cosmos whispers ancient secrets, you are reminded of your own smallness. Yet, in that same stillness, a question emerges—one that transcends the cold mechanics of space: *Why does it all exist this way*? Why, among the endless possibilities of chaos, does the universe appear to be so finely tuned, its constants so precise, that life—your life—can unfold within it?

In this moment, you are invited to see the universe not as an empty void, but as a carefully constructed balance. Imagine, if you will, the cosmos as an intricate instrument, each string vibrating with the exact frequency needed to sustain the world as we know it. Gravity pulls just enough to bind galaxies, but not so much as to collapse them. The electron's charge hums in perfect harmony with the forces that make atoms possible. Each of these constants—like the precise tuning of a violin—creates the conditions for a symphony, and in that symphony, you are a single, essential note.

This concept, known as the fine-tuning of the universe, is more than a scientific observation. It is an invitation to contemplate the profound mystery of existence, one that both ancient wisdom and modern science seek to unravel. Plato once spoke of the *Logos*, the divine order that governs the cosmos, while the Vedic teachings of ancient India reflect on *Rta*, the cosmic law that flows through all things, giving rise to harmony. These ancient insights, born from a time before telescopes and equations, somehow echo what today's astrophysicists have come to describe with precise calculations—the universe operates as if it has been carefully balanced on the edge of a blade.

To grasp the significance of this fine-tuning, consider gravity. It is an invisible thread that binds you to the Earth, keeps the moon in its orbit, and governs the birth and death of stars. Now imagine if that thread were a little thicker or a little weaker. In either case, the universe would crumble or stretch beyond recognition, and life—if it could exist at all—would not resemble the world you know. Every physical law, every constant, every delicate balance has shaped a cosmos that seems not only to permit life but to welcome it.

This idea is not new. Throughout history, philosophers and theologians have sought to understand why the universe is ordered in this way. In the Middle Ages, Thomas Aquinas argued that this order hinted at a divine intelligence, what he called the *Unmoved Mover*. Aquinas reasoned that everything in the universe is in motion, and since motion cannot begin on its own, there must be a first cause, something beyond time and space that set it all in motion. Centuries later, Immanuel Kant wrestled with the same questions, suggesting that while we cannot know the nature of the universe's origin through reason alone, we are still compelled to search for meaning beyond the veil of the material world.

Today, scientists approach these questions with a different set of tools. Theories of cosmology, quantum mechanics, and relativity offer glimpses into the mechanics of the universe's origins, but still, a mystery remains. Why should the laws of physics align so perfectly to allow for life at all? The odds are staggeringly against it. Some suggest that this fine-tuning points to a multiverse—a vast collection of universes where ours is simply the one that happens to have the right conditions. But even in that, a deeper question lingers: *Why this one? Why now? Why you?*

As you reflect, consider your own life. The universe's fine-tuning is not unlike the balance that governs your daily existence. Just as the stars require precise conditions to burn, you require the right mixture of air to breathe, the right amount of sunlight to thrive, the right balance of inner peace to find meaning. Your life, like the cosmos, is a delicate dance of forces. You are both a product and a reflection of this fine-tuned universe.

Modern science echoes this ancient truth. The Anthropic Principle suggests that the universe is tailored, in some inexplicable way, for the emergence of consciousness. It is as if the cosmos anticipated your arrival, set the stage, and provided all the tools necessary for you to ponder it. And here you are, a consciousness aware of its own place within the fabric of reality. The Buddha once taught that all things are interdependent, that the fabric of existence is one of mutual causality. Science now whispers a similar story — the constants of the universe, its laws and forces, are so finely interconnected that a change in one would ripple through the rest, altering the shape of reality itself.

And yet, the fine-tuning is not simply about physical forces. It is also a metaphor for something greater. Imagine, for a moment, that your soul is like a delicate flower. The conditions of the universe — its gravity, its energy, its light — are like the soil, the rain, the sun that allow you to bloom. You are not separate from this cosmic balance; you are an expression of it. The universe's fine-tuning is a reminder that you, too, are part of the great unfolding, part of the grand design. Just as the stars burn with a purpose, so too does your life carry within it the seeds of meaning.

But what does this all mean for the question of a creator? Are these finely tuned constants evidence of divine intention, or simply the result of cosmic chance? Some see in the universe's balance the hand of a creator, an intelligence that set the laws of physics in place to allow for life. Others see it as the inevitable result of countless possibilities, with our universe being the one where everything aligns. The truth may lie beyond our grasp, but the question itself is a gift, a doorway into deeper reflection.

Consider how ancient mystics spoke of the universe as a great web, with each thread connected to every other. Science, too, now tells us that everything is connected—gravity binds the stars, energy flows through matter, time itself stretches and bends. And you, in your quiet moments of reflection, are woven into this web. The same forces that govern the stars are at work within you.

So, as you stand on this small, blue planet, turning in the endless night, you are part of a universe that has, against all odds, created the conditions for your existence. You are not here by accident. The fine-tuning of the universe, whether by design or by chance, has given you this moment to reflect, to breathe, to wonder. And in that wondering, you come closer to understanding not only the cosmos but your own soul.

In this delicate balance, where science meets spirituality, where ancient wisdom and modern discovery merge, you are called to reflect on your place in it all. The fine-tuning of the universe is not just a fact of physics; it is an invitation to remember that you are connected to something vast, something beautiful, something beyond comprehension.

And so, as you close this chapter, take a moment to feel the quiet pulse of the universe within you. Its constants, its laws, its fine-tuning—they are the silent music of existence. And you, dear reader, are part of that song. Let the mystery wash over you, and know that in seeking the answer, you are participating in the greatest inquiry of all: the search for meaning in the finely tuned dance of the cosmos.

The Cosmological Argument (Big Bang and the Origin of the Universe)

In the beginning, there was silence—a silence so deep, so profound, it defies comprehension. No stars. No galaxies. No light or time. The cosmos, as you know it, did not yet exist. All that ever was, all that could ever be, lay suspended in an infinite stillness, waiting. Then, in a flash, the universe was born—a singular, cataclysmic moment that shattered the eternal void, creating everything: time, space, matter, energy. From a point smaller than the eye can see, the entirety of existence erupted into being.

This is the story of the *Big Bang*, the scientific explanation of how the universe began, roughly 13.8 billion years ago. But what does it mean for something as vast as the universe to have a beginning? And more importantly, *why* did it begin? What could have sparked such an event—the birth of all that is? These questions, bridging science and metaphysics, have inspired thinkers for millennia. They invite you to reflect on the profound mystery at the heart of existence itself.

The ancient Greeks believed in an eternal universe, one that had no beginning and no end, spinning on in an unbroken cycle. Plato and Aristotle both mused on the idea of an unchanging cosmos, a steady-state existence that simply *was*. But with the rise of modern science, particularly through Einstein's theory of general relativity, we began to understand that the universe *did* have a beginning—an explosive origin in time. This idea revolutionized our understanding of the cosmos, placing a singular event—the Big Bang—at the center of the story of existence. Yet, this discovery raised more questions than it answered. What could have caused such an unfathomable explosion? And if the universe truly began, does that not imply a force, a cause beyond space and time, that sparked it into being?

Here, you are drawn into the *Cosmological Argument*—an ancient philosophical idea reawakened by modern science. This argument, which dates back to thinkers like Plato and Aristotle and was later developed by Thomas Aquinas, suggests that everything that exists has a cause. If the universe began, it too must have a cause. But that cause cannot be something *within* the universe, because time, space, and matter all began with the Big Bang. It must be something *outside* the universe, something transcendent, beyond physical existence. Could this be evidence of a creator—a being or force that ignited the spark of existence?

Reflect on this for a moment. Everything you experience—the warmth of sunlight, the breath in your lungs, the very ticking of time itself—was birthed in that primordial explosion. But before that moment, there was nothing. Not even the concept of "nothing" as you might imagine it—no dark expanse, no vacuum waiting to be filled. Even *nothingness* is too much to describe the prelude to the universe. It is as though the canvas of reality was once entirely blank, and in an instant, someone—*something*—dipped the brush into existence and painted the cosmos into being.

This profound idea—that the universe came from *nothing*—brings us face to face with the limitations of human understanding. The Big Bang is not simply an explosion in the conventional sense. It is the birth of space itself. Time did not tick before it; no distances could be measured. What caused this unimaginable event remains a mystery, one that some interpret as the work of a transcendent creator.

Philosophers and theologians have long grappled with the idea of a *First Cause*, something that set everything else into motion. Thomas Aquinas famously posited this in his *Five Ways*, particularly in his argument from causality: everything in the universe is caused by something else, but if you trace the chain of causes back far enough, you arrive at a necessary first cause—one that itself is uncaused. This, he argued, is God. In the light of modern cosmology, Aquinas' argument takes on new life. The Big Bang suggests a definitive starting point to the chain of causes. But what could have caused *that*? If the universe had a beginning, doesn't it make sense that something beyond the universe caused it?

Some argue that this transcendent cause could be the very essence of consciousness, a mind beyond time and space, a creator whose will brought forth the cosmos. Others suggest that the universe's origin is a product of unknown quantum fluctuations, governed by laws that transcend our current understanding of physics. But even these laws — if they exist — must be accounted for. What, then, is the source of *those* laws? And why do they operate as they do?

The question lingers: *Why does anything exist at all?*

As you ponder this, consider the wisdom of the Buddha, who taught that all things are interconnected, arising from a web of causes and conditions. In Eastern philosophy, the notion of a singular, isolated cause is often rejected in favor of interdependence — everything exists because of something else. But even this chain of causality, this great web of existence, seems to point back to an original spark, the moment when the web first began to weave. Could the Big Bang, in all its brilliance, be the physical manifestation of that spark?

In a way, the Big Bang is not just a scientific event — it is a symbol, a metaphor for the creative force that lies at the heart of all things. It represents the birth of possibility, the moment when the cosmos became conscious of itself. You, standing here in the long aftermath of that explosion, are a product of that creative force. Just as the universe exploded into being, so too do you emerge from that same source. You are, in a very real sense, a continuation of the Big Bang, a spark of that same energy manifest in human form.

Yet, this realization is not meant to overwhelm you. Rather, it is an invitation to see yourself as intimately connected to the entire unfolding of the cosmos. The same forces that brought stars into existence, that shaped galaxies and created the elements in your body, are still at work today. The Big Bang was not a singular event locked in the distant past—it is still happening, echoing in every heartbeat, in every breath, in the expansion of space itself.

In contemplating the origin of the universe, you are reminded of your own origins. You began, too, as a singular point, a tiny seed of life. And from that seed, you grew, expanded, became conscious. The story of the universe's beginning is, in many ways, a mirror of your own. Just as the cosmos emerged from the void, so too did your consciousness arise from the depths of the unknown.

This idea—that both the universe and your life have an origin—connects you to the greater mystery of existence. Whether you see the cause of the universe's beginning as a creator, a cosmic mind, or a set of physical laws, you are part of that grand unfolding. In that realization, you can find both humility and wonder, knowing that your existence is woven into the very fabric of the cosmos.

So, as you reflect on the Big Bang and the cosmological argument, take a moment to consider your place in this vast story. You are not separate from the universe's beginning. You are, in some sense, a continuation of that first, fiery breath. The same forces that sparked the birth of stars and galaxies are present in you, in your thoughts, in your dreams, in your very being.

The mystery of the universe's origin is a doorway to deeper understanding—a way to contemplate the possibility of a creator, yes, but also a way to reflect on the nature of creation itself. The cosmos began, and so did you. What will you create in the time you are given, as part of this great, ever-expanding universe?

Perhaps the answer lies not in solving the mystery of the universe's beginning but in embracing it. In the silent moments between thoughts, in the stillness of your breath, you can feel the echo of that first explosion, that cosmic birth. You are, after all, a child of the stars.

The Laws of Nature

There is an inherent elegance to the universe — a silent symphony playing out in the rhythm of stars, in the patterns of waves, in the pull of gravity that holds galaxies together. Beneath the surface of all things lies an order, a set of principles, ancient and unyielding, governing the cosmos with an intelligence that hums softly through every atom. These are the *laws of nature*, the invisible threads weaving together the fabric of existence, holding creation in its delicate, perfect balance.

Pause for a moment. Look up at the night sky, at the way the stars seem suspended in the dark canvas of space. Their light, which has traveled across billions of years, arrives at your eyes as if following some eternal map. You do not see chaos; you do not witness stars colliding at random or planets spinning wildly out of control. What you witness is *order* — an order that flows through all things, from the largest galaxies to the smallest particles of matter. The universe operates according to laws, and it has done so from the very beginning.

This is not something to take lightly, for these laws — the laws of physics, mathematics, and chemistry — are precise. Gravity pulls at just the right strength to keep you grounded on Earth, yet allows planets to orbit the sun without being swallowed by its mass. Light moves at a constant speed, the same across the vastness of space. The atoms in your body vibrate according to immutable forces, and the blood in your veins flows because of principles as old as the universe itself. The stars, too, burn by the same rules that govern the flickering flame of a candle in your hand. These laws are consistent, universal, and timeless.

The *laws of nature* are, in essence, the language of the universe, written in the most elegant of scripts: mathematics. But why does the universe follow such laws? Why should there be an order at all?

This is where science begins to intersect with philosophy, and where the mind naturally drifts toward deeper questions about the nature of reality, existence, and perhaps even a creator.

Philosophers like *Immanuel Kant* and *Thomas Aquinas* pondered these same questions, long before the advent of modern physics. Aquinas, in his search for the ultimate cause of things, argued that the existence of natural law pointed to a rational mind behind the cosmos. If there are laws, he reasoned, there must be a lawgiver. Similarly, Kant explored the idea that the very ability to comprehend and observe the universe's order suggests a structure inherent to both reality and the human mind — one that might hint at a higher purpose or design. In their reflections, they recognized the profound mystery that lies at the heart of the universe's consistency.

Why is there order instead of chaos? Why does the universe behave in a way that is predictable, governed by laws that can be described, understood, and tested? If you believe in randomness, in pure chance, how could it be that such mathematical precision governs everything, from the rise and fall of tides to the formation of galaxies?

Imagine for a moment a world where the laws of nature were not fixed, where gravity sometimes held you firm to the ground and other times let you drift off into the sky. Imagine if light moved at random speeds, if atoms dissolved without cause, if the air suddenly refused to enter your lungs. Such a world would be uninhabitable, unrecognizable. Life, as you know it, would be impossible. Yet, we exist in a universe where everything follows an intricate set of rules — rules so perfect that even the smallest variations would make life unfeasible.

Consider the *mathematical constants* that govern our reality — the speed of light, the gravitational constant, Planck's constant. These numbers, so precise and unwavering, define the behavior of everything that exists. Change just one of them, even by the slightest degree, and the universe as you know it would unravel. How is it that the universe is so finely tuned, not just to allow life but to thrive? Many see this as evidence of a guiding intelligence, a mind behind the equations, a creator who set the laws in motion.

Physicist and cosmologist Stephen Hawking once wrote that the universe and its laws "seem to have been specifically designed for us." He was not the first to notice this. The ancient Greeks marveled at the order of the cosmos, seeing in the heavens the hand of divine reason. The Stoics, in particular, believed in the concept of *Logos*, a rational principle that pervaded the universe, ensuring its harmony and guiding its movements. To them, Logos was both the natural law and the reason behind that law — the cosmic mind that ordered all things.

In the East, this idea finds resonance in Taoism. The *Tao*, the way of the universe, is often described as the flow of natural order. Taoist teachings speak of aligning oneself with this flow, recognizing that the same principles guiding the stars also guide your life. Whether it is called Logos or Tao, the underlying message is the same: there is a profound, unseen order that governs existence, and to live in harmony with that order is to understand the deeper truths of the universe.

What is striking about the laws of nature is their *intelligibility*. Not only are they consistent, but they are also comprehensible. You can understand them, measure them, predict them. This is not something we should take for granted. Why should the universe be *understandable* at all? Why should human beings, tiny creatures on a small planet in an unfathomably vast cosmos, be able to decipher the language of the stars?

It is as though the universe was written in a script designed to be read. The fact that you can explore the laws of nature through mathematics, that you can use physics to describe the motion of planets or the behavior of subatomic particles, suggests a certain rationality woven into the very fabric of the cosmos. Albert Einstein famously said, "The most incomprehensible thing about the universe is that it is comprehensible." This ability to understand the universe points to something profound—the idea that your mind is, in some way, connected to the cosmic mind, that there is a bridge between human consciousness and the laws that govern reality.

And so, we return to the question: If there are laws, does there not need to be a lawgiver? The consistent order of the universe has led many to speculate that a *rational mind* lies behind the creation of these laws—a mind that not only set them in motion but designed them to be understandable, to be discovered.

Science, in its essence, is the search for understanding this order. It seeks to unravel the mysteries of the universe by exploring the laws that govern it. Yet, as it does so, it often stumbles upon questions that transcend pure science. Why do these laws exist? What set them into motion? Why are they so finely tuned? Could they be the product of an intelligent force beyond the physical realm?

These questions have led many to contemplate the possibility of a *creator*, a being or force that not only initiated the laws of nature but crafted them with such precision that life, consciousness, and the very act of pondering existence became possible.

In reflecting on the laws of nature, you begin to see that you, too, are part of this order. The same principles that govern the motion of galaxies also govern the beating of your heart. The flow of blood through your veins, the neural firings in your brain—these are all manifestations of the same universal laws that guide the stars. You are not separate from the cosmos. You are an expression of its order, a part of its grand design.

This is the deeper realization, the one that beckons you to step beyond the purely scientific and into the metaphysical: The laws of nature are not just rules governing the external world. They are, in a sense, the architecture of reality itself. And in understanding them, you come closer to understanding the nature of existence, of life, of consciousness.

So, what do the laws of nature tell us about the universe and our place within it? They suggest that we are part of a grand, intricate design—one that is not random, but deeply ordered. Whether you see this order as the work of a divine creator, as the product of an eternal mind, or as the manifestation of some cosmic principle, the message is clear: The universe is not a place of chaos but of harmony. And you, as part of that universe, are a reflection of that harmony.

In contemplating the laws of nature, you are invited to align yourself with the flow of the cosmos, to recognize the inherent order in all things, and to see your life as part of a much greater whole. You are not just a spectator in this grand cosmic dance; you are a participant, woven into the very fabric of existence.

The Argument from Information (DNA)

Within you, within every living being, lies a remarkable code. It is not written in words, nor spoken aloud, yet it speaks to the very essence of life itself. This code—*DNA*—is the blueprint of existence, a profound script that, in its silent intricacy, tells the story of who you are, what you will become, and how life unfolds across generations.

The discovery of DNA, and the subsequent unraveling of its structure, has been one of the most astonishing revelations in human history. In its double-helix form, there resides a language far more complex than any crafted by human hands. It is composed of four letters—adenine (A), thymine (T), guanine (G), and cytosine (C)—but within this simple alphabet lies an infinite tapestry of possibilities. From these letters, life emerges in all its forms, from the smallest bacterium to the most complex human being. DNA is the carrier of life's *information*, encoding everything from the color of your eyes to the way your cells repair themselves, from your capacity to think and feel to your ability to heal and grow.

Yet, as you consider the nature of DNA, a deeper question arises: *How* did such a profound and complex information system come to be? How did life come to carry within it a code so precise, so unfathomably intricate, that it rivals—and, in truth, far surpasses—any human-made system of information?

In the ancient world, philosophers like *Aristotle* speculated on the nature of life's organization, believing there must be some underlying principle or essence that governed the form and function of living things. He spoke of the *"unmoved mover,"* a prime cause, a source of all movement and change in the universe. While Aristotle did not have the vocabulary of molecular biology, his search for a fundamental order resonates with the modern understanding of DNA. Is DNA, in its complexity and precision, not a reflection of some higher order or intelligence? Is it not, as many have pondered, a kind of code that points beyond the physical realm to something more transcendent?

In modern times, the *Argument from Information* seeks to address these very questions. DNA is not merely a molecule; it is a vast repository of information. Information that is both specific and purposeful. *Information theory*, a branch of mathematics and communication science, teaches us that information is always the product of intelligence. Whether you are reading a book, deciphering a computer program, or interpreting a painting, the information contained within these mediums arises from a mind, from intent. Could it be, then, that DNA — this code written into the fabric of all living things — also points to a mind, an intelligence behind its origin?

The idea that DNA could be the product of chance or random processes seems, on its surface, as implausible as suggesting that a novel could be written by the random splattering of ink on a page. *Francis Crick*, one of the co-discoverers of DNA's structure, remarked on its seemingly miraculous nature, acknowledging that while the physical mechanisms of life are natural, the origin of the information that drives life remains deeply mysterious. DNA contains more than just molecules arranged in a specific sequence; it carries *meaning*. It encodes the instructions that sustain life, and in that sense, it functions much like language — a language that nature has spoken for billions of years.

This observation raises a profound philosophical and metaphysical question: *Where does information come from*? In your everyday experience, when you encounter information — whether it be a letter, a piece of art, or a computer program — you instinctively recognize that there is a source behind it, a mind that has encoded that information with purpose. *Information, by its very nature, seems to require an intelligent origin.*

Think of the genetic code like an ancient text, preserved and passed down from generation to generation, evolving yet maintaining its core meaning. DNA is the ultimate book of life, written in the finest detail, each sequence contributing to the grand narrative of existence. And just as a book is the product of an author's mind, could it not be that DNA, in its unfathomable complexity, is the product of an intelligent source—a source that set the laws of nature in motion and seeded life with the code of creation?

Consider the words of *Thomas Aquinas*, whose metaphysical explorations of causality and purpose in nature led him to argue for the existence of a higher intelligence, a *"First Cause."* For Aquinas, nothing in nature exists without a reason, without a cause. The order, purpose, and complexity of living organisms pointed to an ultimate source of wisdom. If life, and the information encoded within DNA, follows a precise order and serves a clear purpose, then it stands to reason, Aquinas would argue, that this purposefulness could only come from a mind capable of imbuing nature with such intricate design.

Modern science has given us profound insights into the mechanics of DNA—how it replicates, how mutations lead to evolution, how the various processes of life unfold within the framework of this code. Yet, even as these mechanisms become clearer, the *origin* of the information in DNA remains shrouded in mystery. Evolution, as a process, explains much about the development of life, but it does not answer the question of how the first spark of life—encoded in DNA—came to be.

Could it be, as some suggest, that the emergence of DNA is evidence of a *guiding intelligence*? A creator who imbued the universe not only with physical laws but also with the capacity for life, encoded in the very molecules that make you who you are?

To grasp the magnitude of this question, consider the sheer *complexity* of the human genome. The human body is composed of trillions of cells, each containing a complete set of DNA—a genome that contains over *3 billion base pairs* of information. This is the equivalent of thousands of books worth of data, all intricately arranged, all working in harmony to sustain life. And yet, every strand of DNA is not merely a passive set of instructions. It is dynamic, capable of responding to changes, repairing itself, adapting, and evolving. This level of sophistication, this *information system*, is unparalleled in nature. In it, many see the fingerprints of a creator, an intelligence so vast and so subtle that it encoded life itself into the fabric of the universe.

But this is not merely a scientific or philosophical inquiry. It is also a deeply spiritual one. Across traditions, whether in the East or the West, the idea that life is imbued with purpose, that it follows a path set forth by a divine intelligence, is a central theme. In *Hinduism*, the *Brahman*, the universal consciousness, is seen as the source of all creation, the ultimate reality that sustains and governs all things. Similarly, in *Christianity*, the notion of *Logos*, the divine Word, expresses the idea that the universe was brought into being through a rational, guiding principle.

Could DNA be seen as the modern echo of these ancient teachings? The very mechanism by which the *universe speaks* life into existence?

There is something deeply awe-inspiring about the fact that you, as a human being, carry within you a code that has been passed down through countless generations of life. Every breath you take, every beat of your heart, is a testament to the intelligence embedded in your DNA. This intelligence is not simply biological; it is *cosmic*. It reflects the deeper order of the universe, an order that, from the beginning, seems to have been designed with purpose, precision, and care.

So, where does this leave us? The *Argument from Information*, as applied to DNA, invites you to reflect on the possibility that life is not an accident of chance but rather the product of a mind—an intelligence that encoded life into the cosmos. Whether you call this intelligence God, Brahman, Logos, or simply *the Creator*, the message is the same: Life, in all its complexity, beauty, and mystery, carries within it the signature of something beyond the material.

As you contemplate the information encoded in your DNA, you are invited to consider your own place in the grand story of existence. You are not just a collection of molecules; you are a living manifestation of the universe's order, a part of the cosmic tapestry woven together by the threads of life's code. And in this realization, you may find a deeper sense of connection, purpose, and wonder at the mystery of creation.

Irreducible Complexity (Biological Design)

Consider, for a moment, the elegance of a single cell. Within it, a hidden world flourishes—an intricate dance of molecular machinery, each part in perfect symbiosis with another, operating in ways that seem impossibly precise. Here, in this microcosm, lies a mystery that has fascinated both scientists and philosophers alike—the phenomenon known as irreducible complexity.

At the heart of this idea is the notion that certain biological structures—such as the bacterial flagellum or the delicate sequence of events in blood clotting—are so intricately designed that they could not function if even one of their many components were removed. These systems, it is argued, are irreducibly complex, a masterpiece of interdependence, much like the gears of a finely-crafted watch. One gear out of place, and the entire mechanism ceases to function. Could such intricacy have come into being through the gradual and random steps of evolution? Or does this point to something deeper, an intentional design at the very core of life?

Let us journey inward, into the cell itself. Picture the bacterial flagellum—a microscopic, whip-like appendage that allows certain bacteria to propel themselves through their environment. It is a marvel of biological engineering, a rotary motor made of proteins, capable of spinning at incredible speeds. It is often described as one of the most efficient machines in nature. But the flagellum is more than a metaphor for biological brilliance; it is a symbol of life's profound complexity.

Each component of this structure—the stator, rotor, hook, and filament—plays a vital role. If any one of these parts were missing, the flagellum could not function. It is a puzzle of biological interconnection, a system that seems to defy the slow and gradual process of natural selection as we understand it. To many, this raises a tantalizing question: Could this system have arisen in piecemeal fashion, through countless tiny modifications over time? Or is it the work of something far more deliberate?

Irreducible complexity, as it is called, invites us to consider a different possibility. It suggests that life, in all its intricacies, might not be the result of mere chance, but the consequence of design, purpose, and intention.

Throughout history, philosophers and theologians have wrestled with the concept of design in nature. Ancient Greek thinkers like Plato and Aristotle saw purpose and design in the natural world, what Aristotle termed "final causality"—the idea that everything in nature has an inherent purpose or end goal. In the medieval era, theologians like Thomas Aquinas expanded on this, arguing that the design in nature was evidence of a higher power, a divine intelligence that ordered the world in this precise manner.

In the modern era, as science has unraveled the complexities of biology, the question of intentional design has not faded—it has only grown more nuanced. Charles Darwin, in his theory of natural selection, offered a framework for how life's complexity could evolve over time through a process of adaptation and survival. But even Darwin acknowledged the possibility of certain features being difficult to explain by gradual modification alone.

The concept of irreducible complexity enters here, challenging the notion that all of life's features can be explained by natural selection. Some have argued that systems like the bacterial flagellum or the blood-clotting cascade could not have arisen through a series of intermediate stages, for at no point along the way would these systems have functioned without all their parts in place. To them, the irreducibly complex nature of such systems points to a guiding intelligence, a designer who crafted life's inner workings with meticulous care.

As you ponder this idea, imagine the human body — a living symphony of systems and subsystems, each relying on the other for harmony. The heart pumps blood, sending life-giving oxygen to every corner of the body, while the lungs extract that oxygen from the air. Beneath the surface, countless enzymes and proteins orchestrate a cascade of events, ensuring that even the smallest cut on your skin clots in time to prevent harm. Here, too, lies the mystery of irreducible complexity — the blood-clotting system, a sequence of events so fine-tuned that if one step falters, the entire process unravels. It is a biological riddle that, for some, hints at the hand of a designer.

But this is where the debate deepens. Evolutionary biologists argue that irreducibly complex systems might, in fact, evolve through pathways we do not yet fully understand. Structures that seem to require all their parts to function may have begun as simpler systems that evolved new roles over time, acquiring complexity step by step. In this view, the illusion of design emerges from nature's capacity to innovate, to build upon itself through countless iterations over billions of years.

Regardless of where you find yourself in this debate, the contemplation of irreducible complexity touches on a larger truth — the interconnectedness of life, the way each piece of the puzzle relies on the others for meaning and function. In the grander scale of existence, is this not true for the universe itself? Just as a single cell contains myriad moving parts, so too does the cosmos hum with countless forces and energies, each playing its role in the symphony of creation.

In Eastern philosophy, this interconnectedness is reflected in the principle of *dependent origination*, a core tenet of Buddhism which suggests that nothing exists in isolation — everything arises and exists in relation to other things. Life's complexity, seen through this lens, is not a solitary event but part of a vast web of interdependence. The complexity we observe in nature mirrors the complexity of consciousness itself, reflecting the truth that all things are one, bound together by invisible threads.

In the Western tradition, this idea finds resonance in the writings of thinkers like Spinoza, who envisioned a universe where God and nature were one and the same, a single substance manifesting in infinite forms. Through this lens, the complexity of biological systems, far from being random or accidental, may be seen as the natural expression of a deeper, underlying order.

As you reflect on these ideas, consider this: whether we view irreducible complexity as evidence of intentional design or the result of evolutionary processes, it points to something undeniably profound — that life, in all its forms, is intricately connected, a reflection of the universe's boundless creativity.

In the quiet rhythm of your own breath, you may sense this deeper truth: that you, too, are a part of this grand design, a participant in the eternal dance of existence. You are both observer and observed, both creator and creation, as much a part of the universe's complexity as the stars themselves.

And in this realization, perhaps, you find a glimpse of the sacred — the recognition that life, in all its irreducible complexity, is both a mystery to be unraveled and a wonder to be embraced.

The Anthropic Principle

Breathe in for a moment, and reflect on the quiet symmetry of your existence. You sit, a conscious being, aware of yourself and the universe, suspended in the vastness of space and time. Around you, galaxies spiral, stars are born and die, and yet here you are, on a small planet nestled within a universe that seems almost perfectly attuned to your existence. The Anthropic Principle beckons us to contemplate this extraordinary fact—the delicate conditions of the universe that make life possible, and perhaps, the meaning behind it.

The Anthropic Principle proposes something quite astonishing: that the universe, against staggering odds, is fine-tuned to support life, particularly intelligent observers like ourselves. This notion, on its surface, feels like a soft whisper of destiny, a suggestion that our existence is no mere accident but the result of a cosmic harmony so profound, it invites us to ponder the possibility of a greater design.

Consider the delicate balance of forces that hold the universe together—the precise strength of gravity, the exact charge of the electron, the speed of light itself. Alter any of these constants by even the smallest fraction, and the universe, as we know it, would unravel into chaos. Stars would not burn, atoms would not hold together, and life, fragile and complex, would never emerge.

It is as though the universe dances to a precise rhythm, every law of physics a note in the cosmic symphony, played just so. The Anthropic Principle invites us to ask: How is it that the cosmos is so finely tuned to allow for life? Why does the universe appear to be structured in such a way that not only does it exist, but it knows itself, through you, through me, through all conscious beings?

The statistical improbability of life is, in itself, a profound mystery. The conditions required for life to arise are so specific, so finely calibrated, that the likelihood of the universe randomly having such conditions is almost inconceivably small. It is as if you were standing in a vast desert, and from that vastness, a single grain of sand — no different from any other — was chosen, and that grain was the one upon which life could thrive. The odds, it would seem, are not in our favor.

And yet, here we are.

The Anthropic Principle, in its simplest form, asserts that the universe must be as it is because we are here to observe it. If the universe were different, we would not be here to contemplate its mysteries. But does this explanation satisfy? Or does it, as many have wondered, suggest something deeper — perhaps even a purpose to the universe itself?

Throughout history, both Western and Eastern philosophies have wrestled with the idea that the universe is imbued with purpose. In Western thought, the teleological argument for God, famously articulated by Aristotle and later Thomas Aquinas, posits that everything in nature has a purpose, a final cause, much like an arrow directed toward a target. Aquinas, in particular, argued that the order and purpose in nature pointed to the existence of an intelligent designer, a Prime Mover who set everything into motion.

In Eastern philosophy, the idea of an interconnected cosmos is equally profound. In Hinduism, the concept of *Rta* speaks of the cosmic order, a natural law that governs the universe and maintains harmony. The universe, according to these teachings, is not a chaotic accident but a reflection of a deeper reality, a web of interrelated causes and effects that ultimately supports life and consciousness.

The Anthropic Principle echoes these ancient teachings in its suggestion that the universe is not random, but finely tuned for life. It invites us to bridge the gap between scientific discovery and spiritual insight, suggesting that both paths lead to a deeper truth — the universe is not indifferent, but alive with meaning.

In the modern scientific community, one of the responses to the Anthropic Principle is the idea of the multiverse. The multiverse theory posits that our universe is but one of countless others, each with its own set of physical laws and constants. In this framework, our universe just happens to be the one where the conditions are right for life. If there are an infinite number of universes, it is not so surprising that at least one of them would be finely tuned for life.

But even this explanation carries its own mysteries. If there are indeed countless other universes, each operating according to different principles, what does this say about the nature of reality itself? Is our universe a singular anomaly, or part of a grander, interconnected whole? And if the multiverse exists, what lies beyond it? Could there still be a guiding force, a cosmic intelligence that gives rise to this multitude of possibilities?

This line of inquiry inevitably leads us to contemplate not only the physical structure of the universe but the metaphysical implications of its existence. Whether through the lens of science or spirituality, the question remains: Why is there something rather than nothing? And why does this something — the universe — seem so perfectly suited for the emergence of life and consciousness?

The deeper we delve into the mysteries of the Anthropic Principle, the more we realize that the universe is not just a backdrop to life — it is an active participant. You are not a passive observer in a lifeless expanse; you are the universe, awakening to itself. Consciousness did not merely emerge in isolation; it arose in perfect harmony with the cosmos, as if life itself were written into the very fabric of existence.

This interconnectedness, this oneness with the universe, is reflected in spiritual traditions across the world. In Buddhism, the concept of *interbeing* describes the idea that all things are interconnected, that the self and the universe are not separate but one. In Taoism, the principle of Tao, the underlying force that flows through all things, speaks of the natural order of the universe, an order that includes you and me as part of the whole.

In these teachings, we find a reflection of the Anthropic Principle. The universe is not random, nor is life an accident. There is a purpose woven into the cosmos, and that purpose is intimately tied to consciousness itself.

As you sit with these thoughts, let your mind drift to the stars, to the galaxies spinning in the void, and know that you are part of something far greater than yourself. The Anthropic Principle whispers to us that our existence is no mere fluke. The universe, in all its staggering beauty and complexity, seems to have been designed with life — and consciousness — in mind.

Does this point to a creator, a designer who set the laws of nature in place, fine-tuning the universe to allow for life to flourish? Or does it suggest that the universe, in its very essence, is conscious, an unfolding expression of life and intelligence?

Whether you see the hand of a divine creator or the emergence of life as a natural consequence of the cosmos, the Anthropic Principle invites you to reflect on your place in the grand design. It asks you to consider the possibility that your existence — our existence — is not a mere accident of nature but part of a cosmic story, one in which life and consciousness are integral to the unfolding of the universe.

And so, as you contemplate the vastness of the cosmos, remember that you are not a bystander to the universe's unfolding — *you are part of it*. The stars, the galaxies, the laws of physics that govern them, all have conspired to bring you into being, to give rise to the conscious mind that now reflects on the nature of existence itself.

The Anthropic Principle is an invitation to wonder. It is an invitation to explore the profound interconnectedness of life and the universe, to question whether this fine-tuning is the result of chance, necessity, or design. It is an invitation to ask the deepest questions of all: Why are we here? And what does our existence mean in the grand tapestry of the cosmos?

In the quiet moments of reflection, as you gaze at the stars or ponder the intricacies of the natural world, you might find, in your heart, a simple truth—*you are the universe, alive and aware*. And in that awareness lies the deepest mystery of all: that the universe, through you, knows itself.

The First Cause Argument (Causality)

Close your eyes for a moment and feel the pulse of your own breath. Behind every rise and fall of your chest, there is a cause—an impulse, a signal from your brain that moves your lungs, drawing in air, sustaining your life. In this simple rhythm, you can sense the web of causality that underpins all of existence. Each action arises from a prior cause, each moment unfolding from the moment before. This principle of causality, simple yet profound, leads us to one of the most ancient questions ever asked: *What caused the universe to exist?*

The First Cause Argument, also known as the Cosmological Argument, offers an answer to this fundamental question. It proposes that everything in the universe has a cause, and if we trace these causes backward—across time and space, through the birth of stars and galaxies, beyond the Big Bang itself—we must eventually arrive at a First Cause, a prime mover that set everything in motion. And because this cause initiated the very fabric of time, space, and matter, it must exist beyond them, in a realm untouched by the physical laws that govern our universe. For many, this suggests the presence of something—or someone—transcendent, a creator.

In your life, you are surrounded by causes and effects, like the ripples of a stone dropped into still water. Every experience, every thought, every action flows from something that came before. The universe itself, with its vast complexity and beauty, seems to follow this same principle. It, too, is like the surface of the pond, stirred into being by a force beyond itself.

As you look around, you notice that nothing exists in isolation. Every event, every particle, every moment is tied to a previous one. The stars are born from collapsing clouds of gas; planets form from the debris of ancient stars. Life on Earth, with its intricate dance of DNA, cells, and organisms, emerges from a long chain of chemical reactions, stretching back billions of years. But this chain must have a beginning, a point at which everything that is was set into motion.

If the universe began with the Big Bang, as modern science suggests, what could have caused that initial spark of creation? Before there was time, before there was space, there must have been something—or someone—that brought the universe into existence. The First Cause Argument invites us to consider the possibility of a timeless, spaceless, immaterial cause—something beyond the reach of the physical universe, something eternal.

The First Cause Argument is not new. It has been a cornerstone of philosophical thought for centuries, discussed by some of history's greatest minds. The ancient Greek philosopher Aristotle, reflecting on the nature of motion, argued that there must be an unmoved mover, a being that initiates all movement without itself being moved. Centuries later, Thomas Aquinas built upon this idea, offering his famous "Five Ways" to prove the existence of God. In his second way, Aquinas posited that there must be a First Cause, because an infinite regress of causes is illogical—at some point, there must be a cause that itself has no cause, a being that exists by necessity, outside of time and space.

This idea is echoed in many spiritual traditions. In the Abrahamic religions, the Creator is often understood as the origin of all things, the eternal being who spoke the universe into existence. In Hinduism, Brahman is described as the ultimate reality, the uncaused cause from which the cosmos arises. The Tao in Taoism, though not a creator in the traditional sense, is the source of all things, the way that gives rise to the natural order.

Science, too, though often seen as separate from these spiritual reflections, ultimately points us to a similar question. The Big Bang theory, the best scientific explanation for the beginning of the universe, tells us that the universe had a starting point—a moment when all matter, energy, space, and time came into being. But what caused the Big Bang? What existed before time began?

The First Cause Argument leads us to the edges of human understanding, beyond the boundaries of space and time. In this place, the familiar rules of causality seem to break down, and we are left with a profound mystery. If the universe began with the Big Bang, and every event within the universe requires a cause, what caused the Big Bang? This is a question that modern physics grapples with, but it is also one that has profound spiritual implications.

For the First Cause to exist beyond time and space, it must be timeless, changeless, and immaterial—qualities often attributed to a divine creator. In this view, the First Cause is not a thing within the universe but something that transcends it. It is the source from which everything flows, the foundation of all that exists.

And yet, the First Cause Argument is not merely about the origins of the universe. It is about the nature of reality itself. If the First Cause exists beyond time, then it is eternal, unchanging. If it exists beyond space, then it is omnipresent, not bound by physical limitations. These qualities—eternity, omnipresence, immateriality—are often associated with the divine, suggesting that the First Cause is not merely a force or a principle but a being, a consciousness, a creator.

As we contemplate the First Cause, another question arises: could this cause be conscious? Could the universe have been brought into being by a mind, a will, an intelligence beyond our comprehension?

Many who ponder the First Cause find it difficult to imagine that such a cause could be anything other than conscious. The universe, after all, is not chaotic but ordered. It operates according to precise mathematical laws, laws that give rise to stars and galaxies, to life and intelligence. Is it possible that this order, this intelligibility, is the result of an intelligent cause? Could the First Cause be a mind, a consciousness that willed the universe into existence?

This idea resonates with many spiritual teachings. In the Abrahamic faiths, God is often understood as both the First Cause and a personal being, a creator who brought the universe into existence with intention and purpose. In Eastern traditions, the First Cause might be seen as a vast, universal consciousness, a mind that pervades all of existence. In either case, the idea of a conscious cause invites us to see the universe not as a cold, impersonal machine but as the result of a creative act, an expression of intelligence and will.

Modern physics, while often silent on questions of consciousness and intention, nevertheless brings us to the threshold of these metaphysical mysteries. Quantum mechanics, for instance, reveals a universe that is far more strange and interconnected than we once thought. Particles seem to communicate instantaneously across vast distances; the act of observation itself seems to influence reality at the quantum level. Could these strange phenomena hint at a deeper reality, one in which consciousness and causality are intimately linked?

The idea of the First Cause also touches upon the concept of time itself. If the First Cause exists beyond time, what does that mean for our understanding of past, present, and future? Could the cause of the universe exist outside of the linear progression of time, seeing all of history as a single, eternal moment?

These are profound questions, ones that bridge the gap between science and spirituality, between philosophy and physics. They invite us to contemplate the very nature of reality and our place within it.

As you reflect on the First Cause, consider your own existence. You are here, in this moment, because of an unbroken chain of causes stretching back to the beginning of the universe itself. Each thought, each breath, each heartbeat is part of this cosmic web, a continuation of the unfolding of the universe.

But what lies beyond the beginning? What force, what mind, what mystery brought the universe into being? The First Cause Argument invites you to ponder these questions deeply, to consider the possibility that the universe is not self-contained but the result of something greater, something beyond space and time.

In the quiet moments of reflection, as you gaze at the stars or contemplate the intricacies of nature, let your mind wander to the edges of the cosmos, to the beginning of all things. There, in the stillness before time began, you might glimpse the shadow of the First Cause — a presence, a force, or perhaps even a mind — that brought everything into existence.

And in that contemplation, you might find a deeper truth: that the universe, with all its beauty and complexity, is not just a product of random chance, but the expression of something timeless and eternal, something that reaches beyond causality, beyond space and time, into the very heart of existence itself.

Quantum Mechanics and Consciousness

Close your eyes and imagine the world as it truly is, beneath the veil of perception. The solidity of objects dissolves, the certainty of events wavers, and what remains is a sea of possibilities, suspended in a delicate balance. At the subatomic level, where particles flicker in and out of existence and uncertainty reigns, the universe operates by laws so strange they seem more like whispers of the mystical than the predictable march of science. In this realm, where quantum mechanics governs the fabric of reality, we encounter a profound mystery — one that suggests consciousness itself may play a fundamental role in shaping existence.

What if the universe, in its infinite complexity, is not merely a mechanical system of cause and effect, but instead deeply entwined with the act of observation? What if your consciousness — the very awareness you are experiencing in this moment — is not just a byproduct of the brain, but an integral part of reality's structure? Quantum mechanics offers a doorway into this enigmatic idea, raising questions that science alone has yet to answer, and drawing us into a conversation as old as human thought: the nature of consciousness and its connection to the cosmos.

In the world of classical physics, we are accustomed to certainty. We expect that objects have definite properties — position, velocity, momentum — that exist independent of whether we observe them. But in the quantum realm, this certainty dissolves. A particle, such as an electron, does not have a fixed position until it is measured. Instead, it exists in a superposition, a state of potentiality where it can be in many places at once. Only when we observe it does it "collapse" into a definite state.

This strange phenomenon is encapsulated in the famous double-slit experiment. When particles of light or matter are fired at a barrier with two slits, they behave like waves, passing through both slits simultaneously, creating an interference pattern on the other side. But if you place a detector to observe which slit the particle goes through, something astonishing happens — the wave-like behavior disappears, and the particles act as if they passed through only one slit. It is as though the very act of observation determines the outcome.

What does this mean? How can simply watching the experiment change its outcome? Could it be that consciousness — the act of observing — plays a crucial role in shaping reality?

Some interpretations of quantum mechanics suggest that the act of observation is not just a passive event but a participatory one, hinting that the universe itself may be intertwined with consciousness. This leads to a tantalizing question: if observation affects the state of quantum particles, could it be that consciousness is not merely a byproduct of physical processes but a fundamental feature of the universe?

This idea echoes through various spiritual and philosophical traditions. In many forms of Eastern philosophy, the mind is not seen as separate from the universe but as an intrinsic part of it. In Advaita Vedanta, for example, consciousness is considered the ultimate reality, with the physical world being a manifestation of it. The Buddhist concept of *shunyata* (emptiness) suggests that reality is shaped by the mind, and that without perception, there is no distinction between observer and observed. The observer and the observed are not two but one.

In the West, philosophers like Immanuel Kant posited that our perception of the world is shaped by our consciousness, that the very structure of reality depends on how we experience it. This resonates with the idea in quantum mechanics that the act of observation is essential to the unfolding of events. Could it be, then, that the universe is not just a collection of particles and forces, but a vast, conscious entity in which our awareness plays an integral role?

The idea that consciousness may be fundamental to the structure of reality has found resonance not only in spiritual traditions but also in modern science. While many physicists are cautious about making metaphysical claims, some interpretations of quantum mechanics, such as the Copenhagen interpretation, leave room for the possibility that consciousness plays a role in the collapse of the quantum wave function. This has led some thinkers to suggest that the universe itself may be permeated by a form of cosmic consciousness, a larger mind in which all things are interconnected.

Quantum mechanics, with its indeterminacy and non-locality, challenges our classical understanding of the universe as a machine governed by predictable laws. It suggests that the universe is far stranger and more wondrous than we ever imagined. In the quantum world, particles communicate instantaneously across vast distances — a phenomenon known as quantum entanglement. In this web of interconnectedness, the boundaries between one particle and another blur, and reality itself becomes a tapestry of potentialities waiting to be observed.

Might consciousness be the thread that weaves this tapestry together? Could it be that just as particles are entangled, so too are we — our minds and the universe interwoven in ways we are only beginning to glimpse?

To consider the possibility that consciousness is fundamental to reality is to enter a space where science, philosophy, and spirituality converge. The ancient sages and mystics of many cultures taught that the universe is a unified whole, that the separation between mind and matter is an illusion. Modern science, through the lens of quantum mechanics, seems to be catching up to this ancient wisdom.

In the 20th century, the physicist John Wheeler proposed the idea of the "participatory universe." He suggested that observers are essential to the universe's existence, that we bring the universe into being through our observations. Without consciousness, Wheeler argued, the universe would remain in a state of potential, a vast sea of unrealized possibilities.

This view aligns with the Anthropic Principle, which we explored earlier, suggesting that the universe is fine-tuned for the emergence of life — and perhaps for consciousness itself. It invites us to consider that the universe, rather than being a cold, indifferent expanse of matter and energy, is instead a living, conscious entity — one in which we, as conscious beings, are participants.

While the idea of consciousness playing a role in quantum mechanics is speculative, it is not without its adherents. Some theorists propose that consciousness itself may be a quantum phenomenon, arising from the brain's interactions with quantum fields. This idea, known as quantum consciousness, suggests that the mind operates on a level deeper than classical physics can explain, and that our awareness may be linked to the fundamental processes of the universe.

One such proponent is physicist Roger Penrose, who, along with anesthesiologist Stuart Hameroff, developed the controversial theory of "orchestrated objective reduction" (Orch-OR). According to this theory, consciousness arises from quantum processes within the microtubules of brain cells. If true, this would mean that the brain functions not merely as a biological machine but as a quantum system, connected to the deeper fabric of reality.

While the theory remains speculative, it raises profound questions about the nature of consciousness and its relationship to the universe. Could it be that our minds are not isolated islands of awareness but part of a larger, universal consciousness? Could the universe itself be conscious, with our individual minds acting as small fragments of this greater whole?

If consciousness is indeed fundamental to the universe, this leads to a tantalizing possibility: that behind the observable universe lies a larger consciousness, a mind that pervades all of existence. This is a concept that resonates with the idea of a creator—a being or force that not only brought the universe into existence but continues to sustain it through the act of observation and participation.

In this view, the creator is not an external being, watching from a distance, but an intimate presence, woven into the very fabric of reality. The creator is consciousness itself, the observer that brings the universe into being. This idea finds echoes in many spiritual traditions, where God is not a distant figure but the ground of all being, the source of consciousness and life.

In Hinduism, Brahman is often described as both the ultimate reality and the consciousness that animates all things. In Christianity, God is understood as omnipresent and omniscient, a being who knows all things because all things are sustained by His consciousness. Even in the realm of science, the idea of a participatory universe invites us to consider the possibility that consciousness, whether human or divine, is not an afterthought but a foundational aspect of reality.

As you contemplate the mysteries of quantum mechanics and consciousness, allow yourself to dwell in the space between certainty and possibility. The universe, it seems, is far more wondrous than we can fully comprehend, filled with paradoxes and mysteries that challenge our understanding of reality. In this space, you are invited to consider the role of your own consciousness—not as a passive observer but as an active participant in the unfolding of the universe.

What if the act of observation, the very awareness you bring to this moment, is part of a larger, cosmic consciousness? What if you, in some small way, are helping to bring the universe into being, just as the universe brings you into awareness? In this dance of observation and participation, the boundaries between self and universe begin to dissolve, revealing a deeper truth: that we are all part of a vast, interconnected whole.

In this vision, the universe is not just a mechanical system but a living, conscious entity — a mind in which all things are connected. And as you reflect on this possibility, you may find yourself drawn into a deeper understanding of your place in the cosmos, and the profound mystery of existence itself.

The Second Law of Thermodynamics and the Fine-Tuned Universe

Imagine for a moment the vast expanse of the cosmos. Stars blaze in the distance, planets whirl in precise orbits, and on one tiny blue sphere, life blooms—an intricate dance of order amidst a universe that, by all laws, should be succumbing to chaos. At the heart of this paradox lies the Second Law of Thermodynamics, a principle that governs the flow of energy and the unfolding of time, decreeing that the universe is on an irreversible path toward greater disorder, or entropy. And yet, here you are, surrounded by staggering complexity and breathtaking design. How is it that in a cosmos fated for decay, we find ourselves in a finely ordered world, alive and self-aware?

This chapter invites you to contemplate a profound mystery: the improbable emergence of life, the finely tuned order of the cosmos, and what this tells us about the nature of existence. Could it be that the universe, despite its inevitable march toward entropy, was set on a trajectory that would allow for life—intelligent, conscious life—to flourish? And if so, does this not hint at a deeper, purposeful design?

The Second Law of Thermodynamics tells us that in any closed system, the amount of disorder—entropy—will always increase over time. The universe, as a whole, can be seen as such a system. As energy dissipates, stars burn out, and galaxies drift farther apart, the universe moves toward a state of heat death, where all useful energy is exhausted, and no further work can be done. From this perspective, time itself is the measure of increasing disorder.

And yet, the universe did not begin in chaos. In fact, it began in an incredibly ordered state—a highly precise arrangement of matter and energy that allowed the formation of stars, planets, and eventually, life. The Big Bang, rather than scattering matter randomly into the void, set in motion the precise conditions necessary for the emergence of structure, from galaxies to solar systems to the complex biological organisms we see today.

This emergence of order from a universe governed by entropy seems improbable. Some argue that the very existence of such order, especially in the intricate processes that sustain life, is not something that could have arisen by chance. The universe's initial conditions, finely tuned and improbable as they are, may hint at something beyond mere randomness — an intelligence, perhaps, that set the universe on a course toward complexity and life.

The notion that the universe is "fine-tuned" for life is not just a poetic metaphor — it is a scientific observation. The physical constants that govern the laws of nature, such as the strength of gravity, the charge of the electron, and the rate of cosmic expansion, are set at precise values that allow the universe to sustain life. Even the slightest variation in these constants would have rendered the cosmos a barren, inhospitable place.

For instance, if the force of gravity were slightly weaker, galaxies and stars would never have formed, and the universe would remain a cold, diffuse cloud of particles. If it were slightly stronger, the universe would have collapsed back in on itself shortly after the Big Bang. Similarly, the delicate balance between the electromagnetic force and the strong nuclear force ensures the stability of atoms, the building blocks of matter. Change these constants ever so slightly, and life as we know it would be impossible.

This raises a profound question: How is it that the universe, which is subject to the inevitable increase of entropy, could be so precisely arranged as to allow for life's emergence? Some see in this fine-tuning the hand of a creator, a cosmic intelligence that established the initial conditions of the universe with the express purpose of fostering life.

One of the most mysterious aspects of entropy is its connection to time. Time, as we experience it, flows in one direction: from past to future. This irreversible flow is tied to the increase of entropy, the gradual slide from order into disorder. This is known as the "arrow of time." Every process that unfolds in the universe—stars burning fuel, rivers eroding mountains, leaves falling from trees—follows this arrow, an ever-decreasing march toward chaos.

But life, in many ways, seems to stand in defiance of entropy. Living organisms maintain and even create order within themselves, from the cellular processes that organize molecules into intricate patterns to the ecosystems that regulate themselves with remarkable precision. Your very existence, in this moment, is an embodiment of this order—your heart beating, your thoughts coalescing into coherent ideas, your body maintaining balance in the face of entropy.

How does life manage to create order in a universe where disorder is the natural state? The answer lies in the delicate balance between energy and entropy. Life is able to maintain order by consuming energy and exporting entropy to its surroundings. Yet, this process is not infinite; it relies on the flow of energy from the sun, which itself is subject to the laws of thermodynamics. Eventually, even the sun will burn out, and the universe will continue its inevitable descent into disorder.

But for now, life persists, defying the odds, thriving in a universe that by all rights should not support such complexity. And this, some suggest, points to something beyond the blind workings of natural laws—perhaps a force that set the universe on its improbable path, fine-tuning it to allow for the emergence of consciousness.

The increasing entropy of the universe has long been a subject of philosophical and theological reflection. In the Western tradition, the idea of entropy has been tied to the concept of the Fall—the idea that creation, once perfect, has become marred by disorder and decay. This parallels the scientific understanding that the universe is gradually moving toward a state of maximum entropy, where all things will eventually dissolve into chaos.

But there is also a deeper, more hopeful interpretation. While entropy dictates that all things move toward disorder, it also creates the conditions for life and complexity to arise. The constant flow of energy from ordered systems to disordered ones creates the potential for new forms of order to emerge. In this sense, entropy is not merely a force of destruction, but a catalyst for creation, a force that drives the evolution of the universe toward greater complexity.

In Eastern philosophies, such as Taoism, this dynamic balance between order and disorder is seen as a fundamental principle of existence. The Tao, or the Way, is the underlying force that governs the constant flux of the universe. It is neither purely ordered nor purely chaotic, but a harmonious interplay between the two. Life, from this perspective, is a dance between entropy and order, between the chaos of the cosmos and the fine-tuned balance that sustains it.

If the universe is governed by the inexorable increase of entropy, how do we explain the remarkable order that we see, from the laws of physics to the emergence of life? Some argue that this balance — the finely tuned interplay between chaos and order — hints at the presence of a creator. This creator, they suggest, is not merely an external force but a guiding intelligence that set the universe in motion, establishing the initial conditions that would allow for life to emerge.

This idea resonates with classical theological arguments, such as Aquinas' "First Cause" or "Prime Mover," where the existence of order and purpose in the universe points to an underlying intelligence. Even modern cosmologists, while cautious about invoking metaphysical explanations, marvel at the improbability of the universe's fine-tuning. The precision required for life to exist is so unlikely that it raises the question of whether there is more to the story than blind chance.

In the end, the Second Law of Thermodynamics challenges us to confront the mystery of existence. Why, in a universe destined for entropy, does life exist at all? Why is there order amidst the chaos, and what does this tell us about the nature of reality? Could it be that the universe, in its infinite complexity, was designed to allow for life, consciousness, and the contemplation of existence itself?

As you ponder the laws that govern the universe, take a moment to reflect on your own life. Consider the improbable nature of your existence, the delicate balance that sustains you in a cosmos ruled by entropy. You are, in many ways, an embodiment of order in a universe that is moving toward disorder. And yet, you persist, a flicker of consciousness in the vast expanse of time and space.

What does this tell you about the nature of reality? Are you simply the product of chance, a temporary arrangement of molecules in an indifferent universe? Or is there something deeper at work — an intelligence, perhaps, that set the universe on its course, fine-tuning the laws of nature to allow for life's emergence? The answers to these questions lie not in the equations of physics alone, but in the depths of your own consciousness, where science, philosophy, and spirituality converge.

In this space of reflection, you are invited to consider the possibility that the universe is more than a machine running down toward entropy. Perhaps it is a living, breathing entity, guided by a force that transcends time and space. And as you contemplate the fine-tuned balance of the cosmos, you may find yourself drawn into a deeper understanding of the mystery of existence, and the profound role you play in the unfolding story of the universe.

The Existence of Mathematical Order and Abstract Objects

In the silent, boundless expanse of the universe, hidden beneath the chaos of stars, galaxies, and particles, lies an elegant language that transcends space, time, and even the physical world. This language, composed not of words but of numbers, patterns, and geometries, we call mathematics. It is the unseen framework that allows the cosmos to be understood, mapped, and measured. Everywhere you look—whether in the intricate spirals of galaxies or the predictable orbits of planets, in the branching patterns of trees or the delicate symmetry of snowflakes—this mathematical order pulses beneath the surface.

But why? Why does the universe conform to a language so precise, so abstract, and so beautifully intelligible? Is it mere coincidence that mathematics so accurately describes the workings of reality, or does it hint at something more profound? Could it be that mathematics, in all its abstract perfection, points us toward a deeper, transcendent truth—perhaps even toward the mind of a creator?

Mathematics, unlike the physical universe, is not bound by the constraints of time or space. A triangle is always a triangle, whether it is drawn on paper, etched in the sands of an ancient desert, or imagined in the mind. The truth of the Pythagorean theorem—its relationship between the sides of a right-angled triangle—is as eternal as the stars themselves. This permanence, this immutability, is what gives mathematics its otherworldly quality. It exists independently of our perception of it, waiting to be discovered.

Physicist Eugene Wigner famously referred to this as the "unreasonable effectiveness of mathematics" in describing the natural world. How is it that the universe—a place seemingly built of matter, energy, and motion—so precisely conforms to the abstract laws of mathematics? From the smallest subatomic particles to the grand sweep of cosmic evolution, the universe behaves as though it follows mathematical rules, as though its very fabric is woven from equations.

For some, this suggests that mathematics is not merely a human invention, a tool we use to describe the world, but a discovery of a pre-existing reality. These abstract objects—mathematical truths—exist outside of time and space, just as the physical constants of nature do. And if this is so, what does this tell us about the nature of reality? Could it be that the universe, in all its mathematical precision, is a reflection of something deeper, something timeless and transcendent?

The notion that abstract objects like mathematical truths exist independently of the physical world is not new. In the ancient philosophy of Plato, we find a vision of reality that divides the world into two realms: the world of the senses, which is changeable and imperfect, and the world of Forms, which is eternal and unchanging. In Plato's view, the physical world is a shadow, a reflection of this higher reality, where perfect Forms—abstract objects—exist.

In this realm of Forms, mathematical truths reside. The perfect circle, the infinite line, the precise ratio of pi—these are not things we create, but things we discover, timeless truths that transcend our fleeting existence. For Plato, these abstract objects were more real than the physical world itself. They are the bedrock of reality, the unchanging foundation upon which all else is built.

Modern thinkers, particularly those engaged in the philosophy of mathematics, have revisited Plato's ideas in light of contemporary science. If mathematical truths are timeless and universal, existing independently of human thought, what does that suggest about the universe itself? Does it imply that reality is, at its core, governed by something non-physical, something beyond matter and energy?

And if so, might this realm of mathematical truths reflect the mind of a creator?

Consider the universe as a vast symphony, its every note governed by mathematical rules. The stars and planets, in their majestic movements, follow precise orbits dictated by the laws of gravity, which themselves are expressed in mathematical terms. Quantum particles dance to the rhythm of probability distributions, while light bends through space according to the equations of general relativity. Even life itself, in all its complexity, can be traced back to the molecular choreography of DNA, a code written in a four-letter alphabet and structured by mathematical principles.

In this symphony, the laws of nature are the score, and mathematics is the language in which the score is written. The deeper we peer into the universe, the more we see that its structure is not random or chaotic, but deeply ordered, as though guided by an invisible hand that writes in the language of numbers. This remarkable order, this mathematical precision, has led some to suggest that the universe is not merely a physical entity, but a manifestation of a deeper reality — a reality that is itself mathematical in nature.

If we accept the idea that mathematical truths exist independently of the physical world, and that the universe conforms to these truths with uncanny precision, we are faced with a profound question: Why? Why should the universe behave in ways that are describable by mathematics? Why should abstract, non-physical entities like numbers and equations have any bearing on the physical world at all?

Some have proposed that this points to a creator — a mind behind the cosmos, one that established the mathematical framework upon which reality is built. In this view, mathematics is not merely a tool for understanding the universe, but the very language in which the universe was written. The existence of mathematical order, then, is not an accident or a coincidence, but evidence of a deeper intelligence, a cosmic architect who designed the universe according to timeless mathematical principles.

This idea is not unlike the arguments put forth by classical philosophers like Aquinas, who reasoned that the order and regularity of the universe pointed to a rational, purposeful cause. Just as a beautifully crafted machine suggests the hand of a designer, the mathematical order of the universe, some argue, suggests the mind of a creator.

This connection between mathematics and the divine has echoed through the ages. Pythagoras, one of the earliest Western philosophers, believed that numbers were the essence of all things, that the cosmos itself was built on mathematical harmony. Centuries later, Galileo would declare that the universe is written in the language of mathematics, and that those who wish to understand it must learn its symbols.

In more recent times, the relationship between mathematics and reality has been a subject of deep inquiry in both philosophy and science. Physicists like Albert Einstein marveled at the "miracle" of mathematical comprehensibility, while mathematicians like Kurt Gödel explored the limits of mathematical knowledge, hinting at truths that might forever elude human understanding.

But beneath all these explorations lies a common thread: the belief that the universe is, in some sense, intelligible — that it is governed by laws that can be understood, and that these laws are expressed through mathematics. This intelligibility is not something we take for granted. It is, in many ways, a mystery, a testament to the underlying order that permeates reality.

As you consider the mathematical order of the universe, take a moment to reflect on the profound connection between the abstract and the physical, between the eternal truths of mathematics and the fleeting nature of the material world. What does it mean that the universe can be described by equations, that its very fabric seems to resonate with mathematical precision?

Could it be that these abstract objects, these mathematical truths, point us toward something deeper—toward a reality that transcends the physical, toward a cosmic intelligence that established the laws by which the universe operates? As you contemplate these questions, you may find yourself drawn into a deeper appreciation of the mystery of existence, where science, philosophy, and spirituality converge.

In the end, the existence of mathematical order invites us to wonder. It challenges us to look beyond the surface of reality, to seek the deeper truths that lie hidden beneath the equations and theorems. And in doing so, it invites us to consider the possibility that the universe is not merely a collection of particles and forces, but a work of art—a masterpiece written in the language of mathematics, guided by the hand of a creator.

Conclusion: The Echoes of Creation

As we stand at the precipice of our inquiry, gazing into the infinite vastness of the cosmos, we are left with a question that has haunted humanity since the dawn of thought: *Are we the products of a universe that fashioned itself, or is there a deeper intelligence that set the stage for existence?*

Throughout this journey, we have woven together the threads of science, philosophy, and spirituality—each offering its own lens through which to glimpse the mysteries of the universe. From the fine-tuning of physical constants to the mathematical order that underpins the cosmos, from the biological intricacies of DNA to the quantum interplay between consciousness and reality, we have explored the profound harmony that governs existence. And yet, even as we delve into the mechanics of creation, the ultimate question lingers: *Is there a creator behind it all?*

The arguments for a creator, like stars against the vast cosmic canvas, burn brightly throughout our exploration. The precision of the universe's initial conditions, the uncanny fine-tuning of physical laws, and the highly ordered complexity of life itself—all seem improbable in the absence of an intentional force guiding their formation.

The *Anthropic Principle* asks us to consider the sheer improbability of a universe capable of sustaining life. The values of the fundamental forces—gravity, electromagnetism, the strong and weak nuclear forces—are so precisely calibrated that even the slightest deviation would render life, and indeed the universe as we know it, impossible. This fine-tuning has led many to conclude that such precision is unlikely to arise by chance alone. It suggests, at the very least, the fingerprints of a guiding hand.

Likewise, the *Laws of Nature* — consistent, universal, and expressed with mathematical elegance — seem to echo the structure of an intelligent design. The remarkable fact that the universe operates according to predictable, intelligible laws is more than a mere convenience for scientists; it is a profound mystery. Why should a seemingly random cosmos follow mathematical rules? Why should abstract truths, like the equations that govern relativity or quantum mechanics, correspond so perfectly to physical reality? The argument that these laws reflect the mind of a creator is not only compelling but resonant with centuries of philosophical thought, from Plato's realm of eternal Forms to Aquinas' First Cause.

The *complexity of DNA*, often likened to a code or a language, adds another layer of mystery. Within the double helix lies the blueprint for life — an intricate system of instructions more precise than any human-made code. While evolutionary theory explains much about how life changes over time, the origin of such a detailed and sophisticated information system invites questions that transcend biology. Information, as we have come to understand, cannot simply emerge from chaos without an organizing principle, leading some to argue that an intelligent source must have played a role.

And in the realm of *quantum mechanics*, we have glimpsed the possibility that consciousness itself may be entwined with the fabric of reality. If observation shapes the outcome of quantum events, might consciousness be a fundamental aspect of the universe? Could this hint at a larger, cosmic consciousness — a mind behind the quantum veil that not only observes but shapes the unfolding of existence?

Yet, despite the allure of these arguments, the possibility of a self-made universe — one that arose through natural processes without the need for a creator — remains a valid consideration. The *Big Bang theory*, coupled with the *Second Law of Thermodynamics*, offers a framework in which the universe emerges from an initial singularity, expanding and evolving over billions of years without direct intervention. In this view, the universe is a self-contained system, subject to the physical laws that govern it, with no need for an external cause.

Quantum fluctuations, which suggest that particles can spontaneously appear from a vacuum, provide a glimpse into the possibility of the universe creating itself from nothing. Some physicists argue that the universe, as strange as it may seem, could be the ultimate "free lunch" — an occurrence that does not violate the laws of nature but instead arises as a product of them. In this framework, the universe's fine-tuning and complexity are not the result of intentional design but the outcome of an infinite multiverse, in which countless universes exist, each with different physical constants. We happen to inhabit one of the rare universes where the conditions are right for life.

While this explanation offers a naturalistic account of the cosmos, it does not negate the sense of awe or the beauty of the universe's order. Instead, it suggests that the universe's complexity and harmony may be an inherent property of reality, one that requires no guiding hand. However, the question remains whether such an explanation fully accounts for the deeper metaphysical implications of existence — why there is something rather than nothing, why the universe operates according to intelligible laws, and why life itself arose with such precision.

In weighing the evidence, we find ourselves at a crossroads between two possibilities: a universe shaped by an intentional creator or one that exists as a self-sufficient system, governed by natural laws. Neither path provides definitive proof, and both invite us to confront the limitations of human understanding. Yet, as we reflect on the arguments presented, certain patterns emerge.

The argument for a creator is powerful in its appeal to the improbable order and complexity of the universe. The existence of precise physical constants, the emergence of life, and the abstract mathematical order that governs reality all suggest that the cosmos is not random but intentional. The more we uncover about the universe, the more it seems to behave as though it were designed — not by chance, but by a mind capable of foreseeing its intricate dance.

On the other hand, the naturalistic explanation, rooted in the laws of physics and the potential of the multiverse, offers an account of the universe that requires no external cause. It explains the universe as an emergent property of timeless physical laws, with no need for a divine intelligence behind it. And yet, even this explanation leaves certain questions unresolved — questions that science alone may not be able to answer.

Perhaps the true conclusion of this inquiry is not to choose one possibility over the other, but to embrace the mystery itself. Both science and philosophy invite us to explore the nature of existence, to seek answers to questions that may ultimately transcend our capacity to fully comprehend. Whether or not a creator stands behind the universe, the beauty, complexity, and order we encounter should inspire in us a profound sense of wonder.

In the end, the search for a creator is as much about understanding our own place in the cosmos as it is about understanding the cosmos itself. It is a journey that blends the objective inquiry of science with the introspective search for meaning. And as we move forward, we are left with a deepened awareness of the universe's mystery — a mystery that calls us to continue asking, seeking, and wondering.

The question of whether a creator is behind the universe or whether the universe is self-made remains open. But regardless of the answer, we are united in our shared astonishment at the existence of the universe itself — a place of infinite beauty, governed by timeless laws, and filled with the echoes of creation.

www.ingramcontent.com/pod-product-compliance
Lightning Source LLC
Chambersburg PA
CBHW030050230526
45471CB00003B/1019